Journey to **MARS**

A SCHOLASTIC /
MADISON PRESS BOOK

Footprints on Mars

This is it! The time for dreaming has ended.

You've wanted to travel to Mars ever since you were a kid. For years you worked to make it happen. And now you're here! In a minute you will plant your feet in Martian soil. And you still can't believe it.

You feel awkward in your hard-shell spacesuit. It has been months since you last wore it. Your helmet is heavy and the visor cuts off what you can see on either side of your head. The chamber you are standing in—the airlock—is short and narrow. Your elbows rub the walls and the ceiling is low. If you jumped just a little you would bang your head. You step carefully towards the exit hatch. You reach for the handle and turn it. The metal pins that have kept the hatch locked for nine long months are released one by one. And then you're ready. All you have to do is push.

You stop for a moment.

You have been living in cramped quarters for so long that you can hardly remember what it's like to stretch out your arms and not hit something. To be outside would be amazing! But the world beyond this door is like no other. No other human has ever landed here. No other human has seen what you are about to see.

You take a deep breath. You stand very still. And then you exhale slowly and push open the hatch.

Red dust—Martian dust!—swirls in through the gap as the door swings open. There are stones on the ground below.

You push the hatch wide open and lift up your head. Mars opens up before you!

J Matthews

From airlock to ladder to the surface of Mars where neither man nor woman has ever set foot. The end of the longest journey and the beginning of the most exciting adventure ever! The photograph below shows one of the first footprints on the Moon, left there more than forty years ago.

The sky above is the color of a ripe peach. The rust-red horizon is jagged and high. The dusty soil shimmers and shifts beneath the spacecraft.

Suddenly you are seized by the crazy wonder of the moment. You cast aside all caution or restraint. You set one foot on the edge of the hatch and swing the other one onto the metal steps. You pause for a second and then leap . . .

Getting here was serious business—but being here is incredibly fun!

For centuries, humans gazed at Mars.

They looked with their naked eyes, saw its blood-like hue and imagined it was the home of the gods of war. They looked through telescopes and saw dark and light patches. They imagined they saw mountains and seas. They launched spacecraft to examine its surface from the sky above it. They dropped robotic vehicles on it. The spacecraft and robots sent back images and data to Earth. Scientists analyzed this information. They learned more about what the planet really was like. They learned enough to send humans to Mars.

The *Troublemaker*

Stars move through the sky in a fixed pattern. Ancient astronomers charted their paths and named the constellations. But the planets were different from the stars. They moved through the skies independently and, it seemed, unpredictably.

Mars, especially, was different. For one thing, it was red. There was something dangerous-seeming about a celestial body that was the color of blood. For another thing, it was all over the place. In some years it was larger, as if closer to the Earth; in other years it was smaller. Its path was erratic.

Because of its color and strange behavior, the ancients associated Mars with trouble and conflict. In Roman religion, there was a cult dedicated to the god Mars. He had his own month, March (*mars* in French). Two-horse chariot races were held in his honor. The right-hand horse of the winning team was sacrificed for him. (Contestants must have tried hard to come in second.) A great temple called Mars Ultor (Mars the Avenger) was built by the emperor Augustus Caesar in 2 B.C. It was considered one of the most magnificent temples in Rome.

In Greek mythology, the god associated with Mars was Ares. While some Greek gods were powerful, others had human-like weaknesses. Ares was relatively unimportant and played only a minor role in their stories. He was a troublemaker rather than a warrior. When he had an affair with Aphrodite, the goddess of love, he was caught in a net and the other gods made fun of him.

Worrier or Warrior? The Roman god Mars was
more farmer than fighter. But in Greek mythology,
Mars' counterpart, Ares, stood for strife.

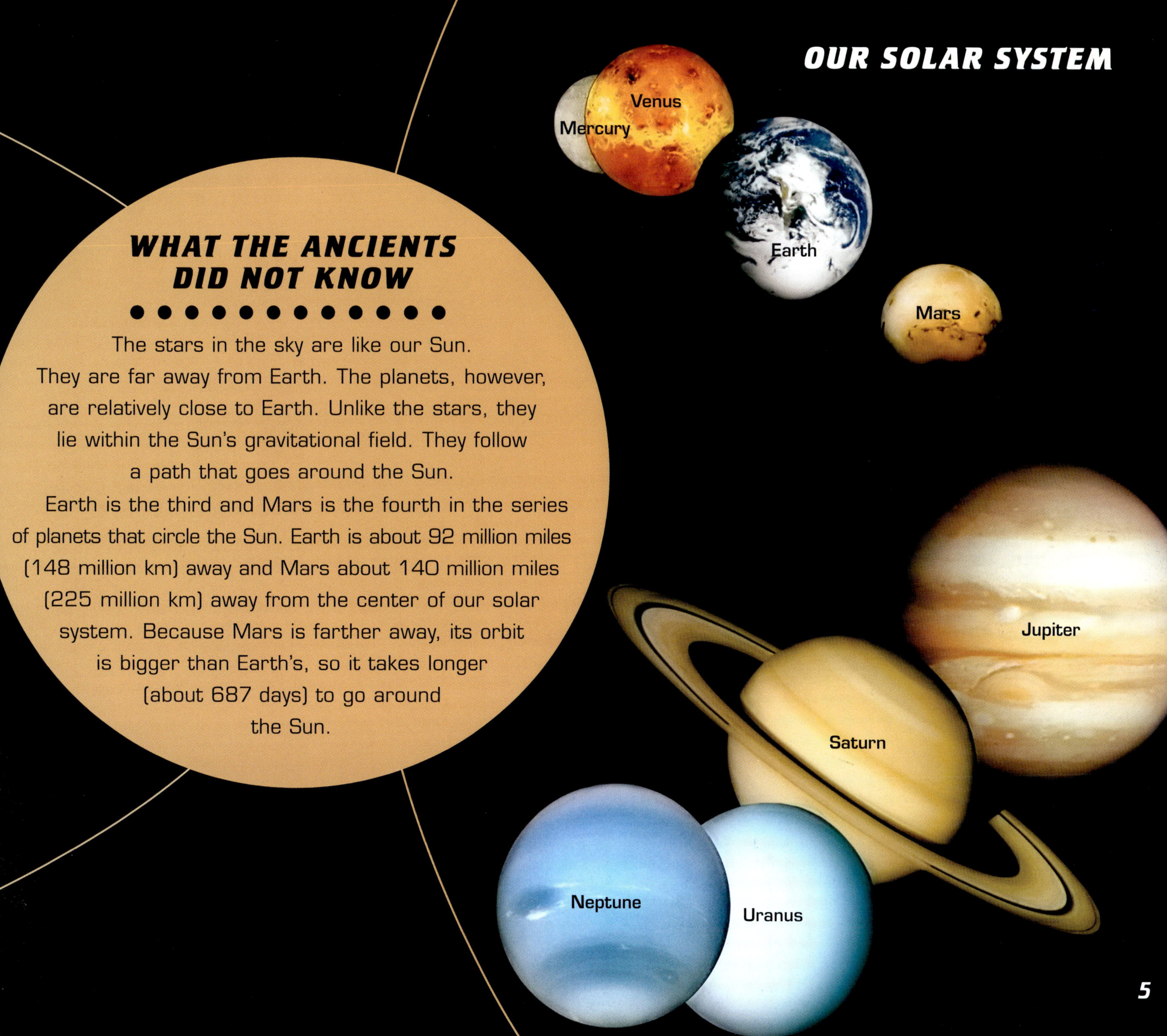

WHAT THE ANCIENTS DID NOT KNOW

• • • • • • • • • • • •

The stars in the sky are like our Sun. They are far away from Earth. The planets, however, are relatively close to Earth. Unlike the stars, they lie within the Sun's gravitational field. They follow a path that goes around the Sun.

Earth is the third and Mars is the fourth in the series of planets that circle the Sun. Earth is about 92 million miles (148 million km) away and Mars about 140 million miles (225 million km) away from the center of our solar system. Because Mars is farther away, its orbit is bigger than Earth's, so it takes longer (about 687 days) to go around the Sun.

Through a Glass, Darkly

Astronomers learned more about Mars at the same time as they learned more about the universe. The Polish astronomer Nicolaus Copernicus figured out that the Earth spins on a north-south axis once every twenty-four hours. This is why we have day and night. He also argued that Earth and the other planets travel in circles around the Sun. Both insights made it easier to understand the changing appearance of Mars.

Copernicus looked up at the night sky with only his eyes to see what was there. So did Tycho Brahe. Brahe, a Danish nobleman famous for his wild parties, made amazingly accurate observations of the planets and stars. He gave his notes to a German mathematician and astronomer, Johannes Kepler. It was Kepler who, looking at Mars, realized that the planet sometimes was closer and sometimes farther away from the Sun. This meant that its path was not circular, but elliptical.

A half century later, the Dutch mathematician Christian Huygens noted that Mars, like the Earth, spins on a north-south axis. He estimated correctly that the Martian day was about as long as Earth's. And a few years later in 1666, Giovanni Cassini, using a more powerful telescope at an observatory in Italy, reported seeing polar ice caps on Mars! Bit by bit, these men were adding new information to what was known about the Red Planet.

Johannes Kepler

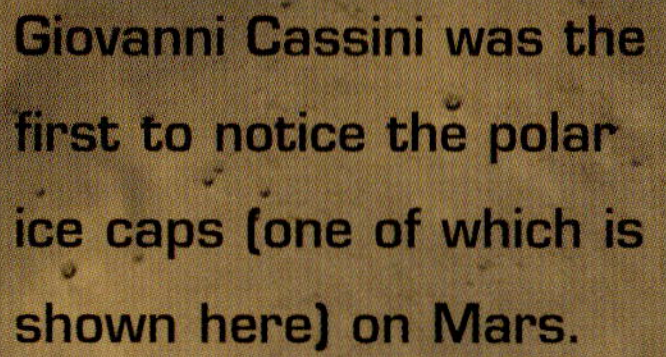

Giovanni Cassini was the first to notice the polar ice caps (one of which is shown here) on Mars.

SEEING IS BELIEVING

In 1609, the Italian scientist Galileo Galilei made his own telescope and pointed it at the skies. It was a crude instrument by modern standards. He wrote to a friend: "I dare not affirm that I am able to observe the phases of Mars; nonetheless, if I am not mistaken, I believe I have seen that it [Mars] is not perfectly round."

GEOGRAPHY, FIRST PERIOD

The early 1800s are sometimes called the first Geographic Period in the history of Mars investigations. Astronomers had better telescopes by this time. They used them to make more detailed observations of the planet. They also started to wonder if there was life on Mars.

The English composer and scientist William Herschel had already noticed that there were light and dark patches on the planet's surface. He thought that the dark patches might be bodies of water.

Two enterprising Germans went further. William Beer (who had a day job as a banker) and Johann H. von Mädler (an astronomer) invented a system of latitude and longitude for Mars similar to the system we use to draw maps of Earth. We still use their system to identify locations on Mars.

Many more maps were made of Mars in the 1800s. Telescopes brought the skies closer to Earth. But not close enough to be really accurate. Their blurry view of Mars led to over one hundred years of mistaken speculation.

Galileo's telescopes

What's in a Name?

Johannes Kepler figured out that the distance between Mars and Earth varied as the two planets followed their separate paths around the Sun. The difference can be drastic. At roughly fifteen-year intervals, the two planets may be as close as 30 million miles (50 million km) and as far apart as 60 million miles (100 million km). When they are at their closest, they are said to be in *opposition*.

An unusually favorable opposition occurred in 1877. Astronomers all over the world took advantage of the occasion to focus their telescopes on Mars. An American, Asaph Hall, used a 26-inch (66-cm) refractor telescope at the U.S. Naval Observatory in Washington to discover that Mars has two moons. And an Italian, Giovanni Schiaparelli, looked up at Mars from the Brera Observatory in Milan and was fascinated by the light and dark areas of the planet to which he gave fanciful names. We still use some of Schiaparelli's Martian place names. For example, we call some regions that he thought of as deserts *hellas* and *amazonis*. Two regions that he thought were seas are still known as *Boreum Mare* and *Syrtis Major*.

Significantly, Schiaparelli reported that the Martian surface appeared to be marked by a network of straight narrow lines, which he called *canali*. *Canali*, as Schiaparelli used the term, meant simply channels, but his choice of the word had momentous consequences.

TAKING A DIM VIEW

When Mars is closest to us, Earth sits between it and the Sun. In effect, Earth blocks the sunlight that would make it easier to see the surface of the Red Planet. So the best time to observe Mars is also the worst!

THE COMING CRASH

Mars has two moons. They are named Phobos (a Greek word meaning *fear* or *dread*) and Deimos (meaning *terror*) after the two horses that pulled Ares' chariot. Both moons are relatively small and oddly shaped. Scientists believe that they are asteroids that have been pulled from space into the Martian gravitational field. Phobos travels on an unstable path and is expected to crash into Mars in several hundred million years.

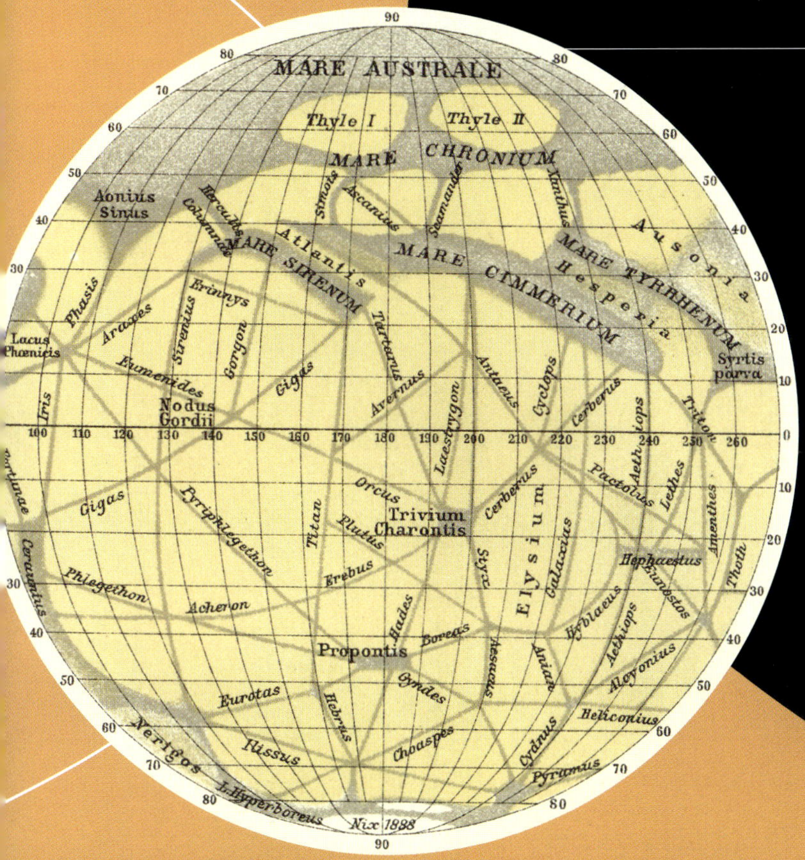

Schiaparelli's map showed lines connecting dark regions on the surface of Mars.

"There are on this planet, traversing the continents, long dark lines which may be designated as canali, although we do not know what they are. Those lines run from one to another of the somber spots that are regarded as seas, and form, over the lighter or continental regions, a well-defined network. Their arrangement appears to be invariable and permanent . . . ."

—Giovanni Schiaparelli

A Dying Civilization

Was there life on Mars? Some were sure there was.

Percival Lowell was a gifted mathematician and scholar who dedicated the final fifteen years of his life to investigating Mars from Earth.

Lowell was among the first scientists to appreciate the importance of viewing the skies in the clearest possible conditions. He established an observatory at Flagstaff, Arizona in 1894. The altitude there is about 7,000 feet (more than 2,000 m) above sea level. The climate is almost always clear and dry. It is a nearly ideal location for an observatory.

Lowell was intrigued by the lines on the maps of Mars drawn by Giovanni Schiaparelli. He was especially taken by the suggestion that the lines might represent *canals*. Beginning in 1894, when Mars was again in opposition, he began drawing maps that he thought would be more accurate than the Italian astronomer's. The more he worked on his maps, the more convinced he became that he was seeing evidence that there was intelligent life on the Red Planet.

Lowell understood that the lines he saw on Mars could not be actual canals. Any line visible from Earth, even through his telescope, had to be bigger than a waterway. He decided instead that they were the cultivated fields on either side of waterways. He reckoned that the fields were irrigated using meltwater carried in canals from the polar ice caps.

Percival Lowell believed there was intelligent life on Mars.

But why were the canals necessary? Lowell believed that the climate of Mars was becoming drier and that the Martians had built the canals to bring water to their fields. He thought that the canals showed that Martian civilization was slowly, tragically dying.

Lowell's books about Mars were tremendously popular. His ideas influenced generations of artists and writers who shared his beliefs.

AN OPTICAL ILLUSION

When Percival Lowell looked through his telescope, he made colored sketches of what he saw. These drawings (see three insets here) show dark patches, splotches, and dots joined together by shaded areas and lines. Scientists who study how people think believe that we always try to make sense of things. Our brains tend to look for a pattern in everything we perceive. If we see dots, we try to connect them to make a picture. Lowell wanted to believe there was intelligent life on Mars. He wanted to see a pattern—and so he did!

A SKEPTIC

An American astronomer, Edward E. Barnard of the Lick Observatory in California, expressed a widespread professional view in a letter to colleagues. "I have been watching and drawing the surface of Mars," he wrote. "It is wonderfully full of detail. There is certainly no question about there being mountains and large, greatly elevated plateaus. But to save my soul, I can't believe in the canals as Schiaparelli draws them. I see details where he draws none. I see details where some of his canals are, but they are not straight lines at all. I verily believe all the verifications that the canals depicted by Schiaparelli are a fallacy . . . ."

Mars Mania!

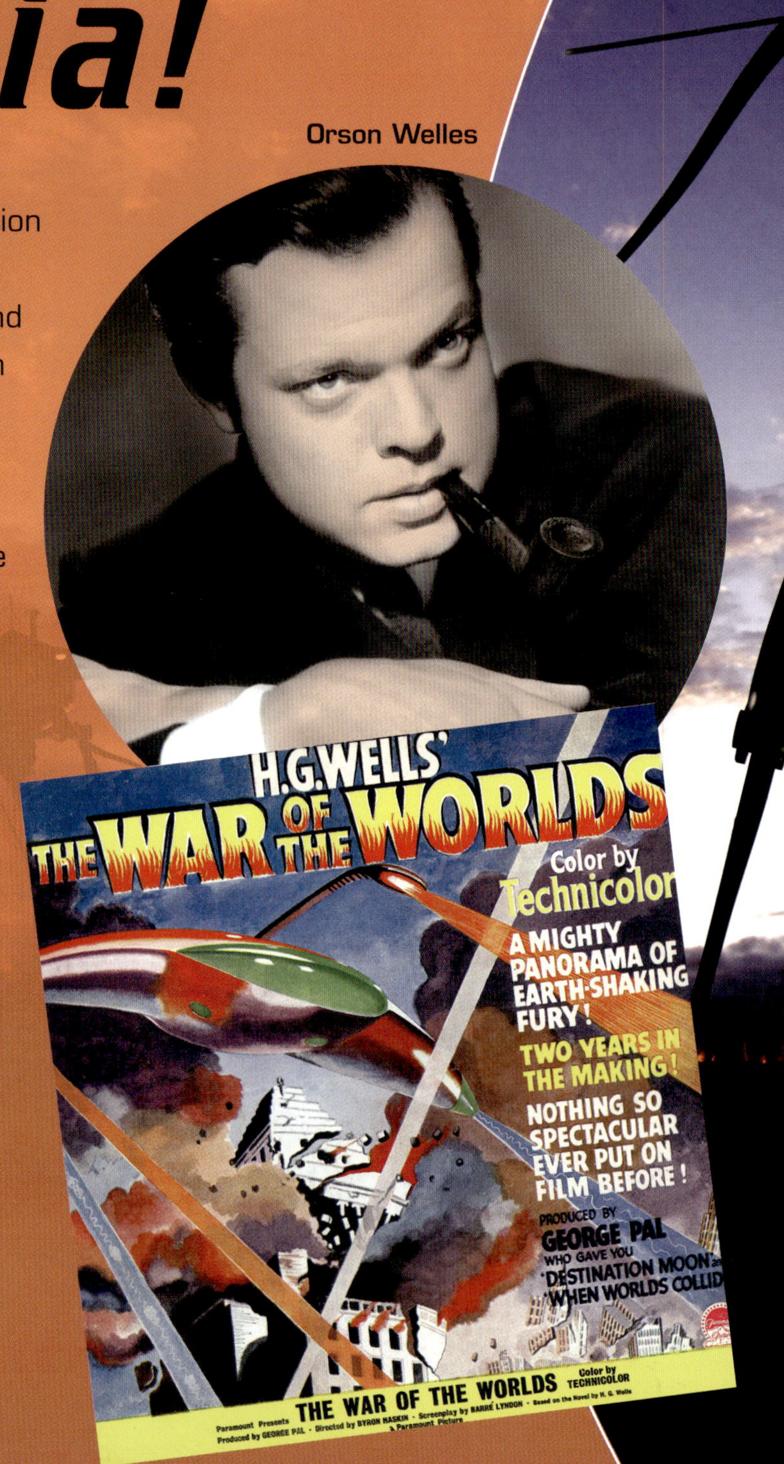

H.G. Wells was one of the earliest writers of science fiction stories. In his novel *The War of the Worlds*, he built on Lowell's idea that Mars was the home of an intelligent and sophisticated race of beings. He accepted Lowell's notion that our closest planetary neighbor was undergoing drastic climate change. Wells, however, suggested that instead of getting drier, Mars was slowly freezing. He took the idea a step further. He wrote that the Martians were abandoning their planet. They wanted to take over Earth.

The *War of the Worlds* has been read by millions of people since it was first published in 1898. Several film versions of the story have been made. And on one famous occasion, in 1938, the story was presented by the actor and director Orson Welles as a radio play in the United States. Actors pretended to interrupt a regular radio program with bulletins about mysterious explosions on Mars. Then they reported that some kind of spaceship had landed in New Jersey. Thousands of listeners believed the reports were true. There was widespread panic as people tried to escape the Martian invasion.

Wells was among the first of many writers whose imagination was stirred by Mars . . .

Edgar Rice Burroughs (the creator of Tarzan) wrote a series of novels in which his hero, John Carter, did battle on Mars with a variety of terrifying aliens.

Two comic strips featuring resourceful heroes in a future time, Buck Rogers and Flash Gordon, were tremendously popular among newspaper readers from the 1930s until the 1960s. Their adventures often took them to Mars.

The novelist Ray Bradbury wrote a series of connected stories about humans escaping Earth to establish a colony on Mars. *The Martian Chronicles* has captivated tens of thousands of readers.

There have been countless other films, books, comics, and even television programs based on the idea that Mars shelters life forms more or less similar to Earth's. Perhaps the silliest was the television series, *My Favorite Martian*, which was broadcast in the 1960s. In the series, a grumpy Martian found himself stranded when his flying saucer crash landed on Earth. He moved in with a young newspaper reporter, pretended to be his uncle, and solved the young man's everyday problems by exercising magical Martian powers.

MAD MARTIANS?

Sometimes we have imagined that Martians are hostile. At other times, we have imagined that they are friendly. But the idea that Martians exist, for better or worse, is a powerful one. Some people just can't give it up. They still believe that there is life on Mars.

Martian Round-About

Between 1965 and 1971 four American spacecraft either flew past or circled Mars. *Mariner 4* and *Mariner 9* were especially significant.

MARINER 4

On July 15, 1965, *Mariner 4* snapped twenty-two sketchy photographs of the southern half of Mars from a distance of about 5965 miles (9,600 km). The grainy, black-and-white pictures were transmitted back to Earth where they came as a shock. Scientists had expected to see a landscape similar to Earth's where wind and water have softened jagged mountains and worn away the sharp edges of canyons and craters. Instead they saw a harsh, pockmarked surface similar to our Moon's.

There was no vegetation. There was no sign of water. Nothing remotely like a canal could be seen anywhere.

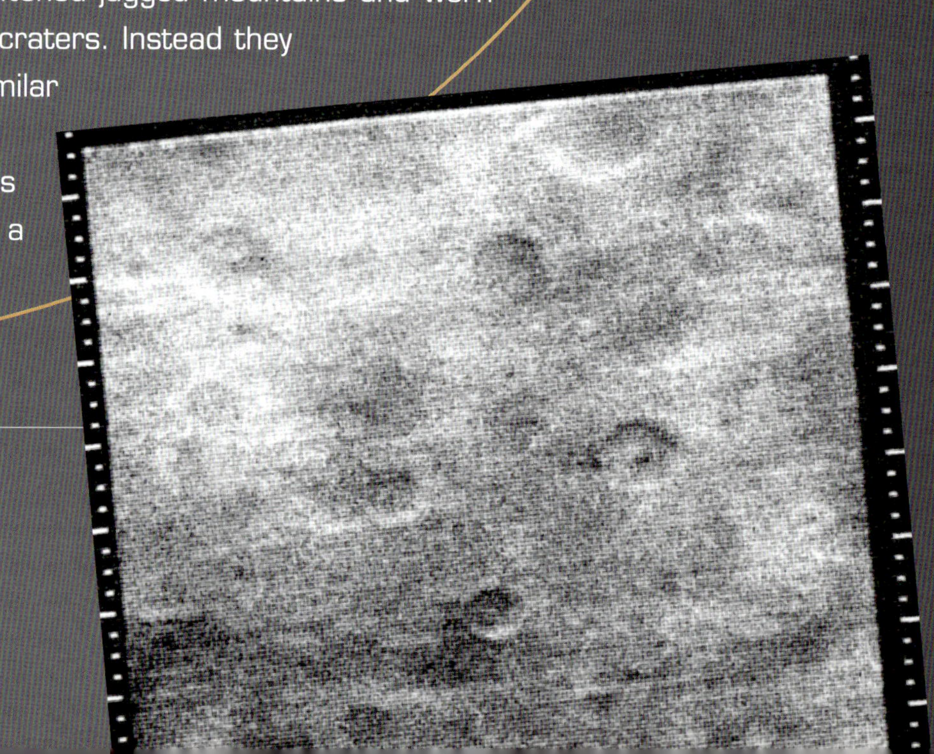

Mariner 4

The first pictures of Mars sent back to Earth by *Mariner 4* showed a cratered moon-like planet. This image was taken from about 8,000 miles (13,000 km) above the surface.

MARINER 9

Earlier probes in the Mariner series flew past Mars. On November 14, 1971, *Mariner 9* entered Mars orbit and stayed there. It became, in effect, the first artificial satellite to orbit another planet.

The pictures transmitted to Earth from *Mariner 4* had been grainy and unclear. *Mariner 9*'s first pictures weren't much better. The spacecraft arrived when a huge dust storm was sweeping across the planet. When the dust settled, however, the view improved considerably. And because *Mariner 9*'s camera was more sophisticated, the pictures it transmitted home were much sharper than *Mariner 4*'s had been. There were many more of them too—about seven thousand. These pictures, in their way, were just as shocking as the earlier ones had been.

Mariner 9 revealed a rugged and varied world. It showed mighty volcanoes, enormous trenches, deep faults, broad river valleys, and meandering channels. The sight of these channels and river valleys made scientists open their eyes in surprise. They could mean only one thing: there once had been water on Mars! There might still be water on Mars. And where there is water, there is the possibility of life.

Mariner 9

Phobos

SPUDS IN SPACE?

• • • • • • • • • •

Mariner 9 snapped the first close-up photographs of the Martian moons. Seen through a telescope from Earth, Phobos and Deimos are mere pinpoints of light. Seen up close, they look like potatoes!

The Mars Monster

Humans have sent thirty-nine missions to Mars over the years. Just seventeen have reached their destination. Thirteen missions were intended to land on Mars. Only six succeeded. The record is brutal. Russian missions have been especially unlucky. Some observers have joked that there must be a monster out there, waiting to devour vital bits of passing spacecraft. Only the existence of such a monster explains the disappearance and failure of so many expeditions. John Cassani, a scientist at the Jet Propulsion Laboratory, called it the "Great Galactic Ghoul."

Mars Polar Lander

What the Monster Ate ...

Year	Mission	Outcome
1960	*Korabl 4* (Russia)	Failed to reach Earth orbit
1960	*Korabl 5* (Russia)	Failed to reach Earth orbit
1962	*Korabl 11* (Russia)	Earth orbit only, spacecraft broke apart
1962	*Mars 1* (Russia)	Radio failed
1962	*Korabl 13* (Russia)	Earth orbit only, spacecraft broke apart
1964	*Mariner 3* (USA)	Mechanical breakdown, shroud failed to jettison
1964	*Zond 2* (Russia)	Radio failed
1969	*Mars 1969A* (Russia)	Failed to reach Earth orbit
1969	*Mars 1969B* (Russia)	Failed to reach Earth orbit
1971	*Mariner 8* (USA)	Launch failure
1971	*Kosmos 419* (Russia)	Earth orbit only
1971	*Mars 2* (Russia)	Lander crashed
1971	*Mars 3 Orbiter/Lander* (Russia)	Lander communications failed after 20 seconds
1973	*Mars 4* (Russia)	Failed to enter Mars orbit
1973	*Mars 6 Orbiter/Lander* (Russia)	Partial success, lander failed on descent
1973	*Mars 7 Lander* (Russia)	Missed planet
1988	*Phobos 1* (Russia)	Failed to reach Mars
1988	*Phobos 2* (Russia)	Lost contact after 2 months
1992	*Mars Observer* (USA)	Lost prior to Mars arrival
1996	*Mars 96* (Russia)	Launch failure
1998	*Mars Climate Orbiter* (USA)	Failed to enter Mars orbit
1999	*Mars Polar Lander* (USA)	Crashed on Mars
2003	*Mars Express Orbiter/Beagle 2 lander* (Britain)	Partial success, lander lost on arrival

Orbiters and Landers

The Mariner probes were a giant step forward on the way to Mars. The *Viking* and *Rover* programs took us even closer.

VIKING 1

The *Vikings* were two spacecraft in one. Each consisted of an orbiter equipped with cameras and sensing instruments and a robot lander whose task it was to analyze the surface. Both *Vikings* entered Mars orbit in the summer of 1976.

Before it could get on with its chief job, *Viking 1* first had to pick a place to land. The scientists who had planned the mission thought they were sending the lander to a flat region. Photographs sent back by the orbiter showed that the intended site, in fact, was in the bottom of a crater. The landing was delayed for more than two weeks while people on the ground examined the photographs sent back from space. They finally picked a spot roughly 480 miles (800 km) away from the original site.

In August 1976, the *Viking 1 Lander* sent back this image of a wide, low Martian plain covered with rocks, sand, and dust.

The lander touched down on July 20, 1976. Scientists were thrilled. The project manager, Jim Martin, said later, "It was all so unbelievable. Here was this lander 200 million miles (322 million km) away, sitting on another planet, and we were on Earth looking at pieces of dirt on its footpad. I felt a little strange... the first picture that mankind had ever seen from the surface of another planet. It was so clear; it was as if we were standing right there beside it."

WHODUNIT?

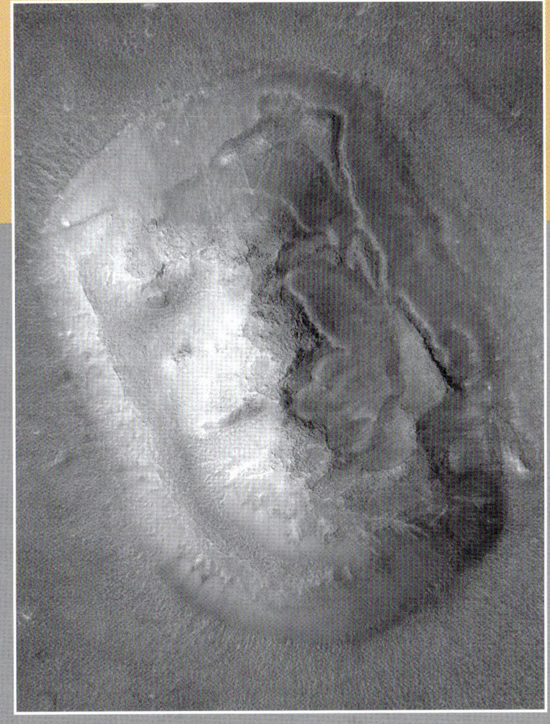

Viking 1 Orbiter sent to Earth one picture in particular (right) that got many people excited. The image appears to show a human-like face sculpted into the surface of the planet. Was a Martian artist trying to tell us something? Well, no, probably not. A more recent, better picture taken from a different angle (below) of the same surface area on Mars reveals that the face is just a bit of rough terrain.

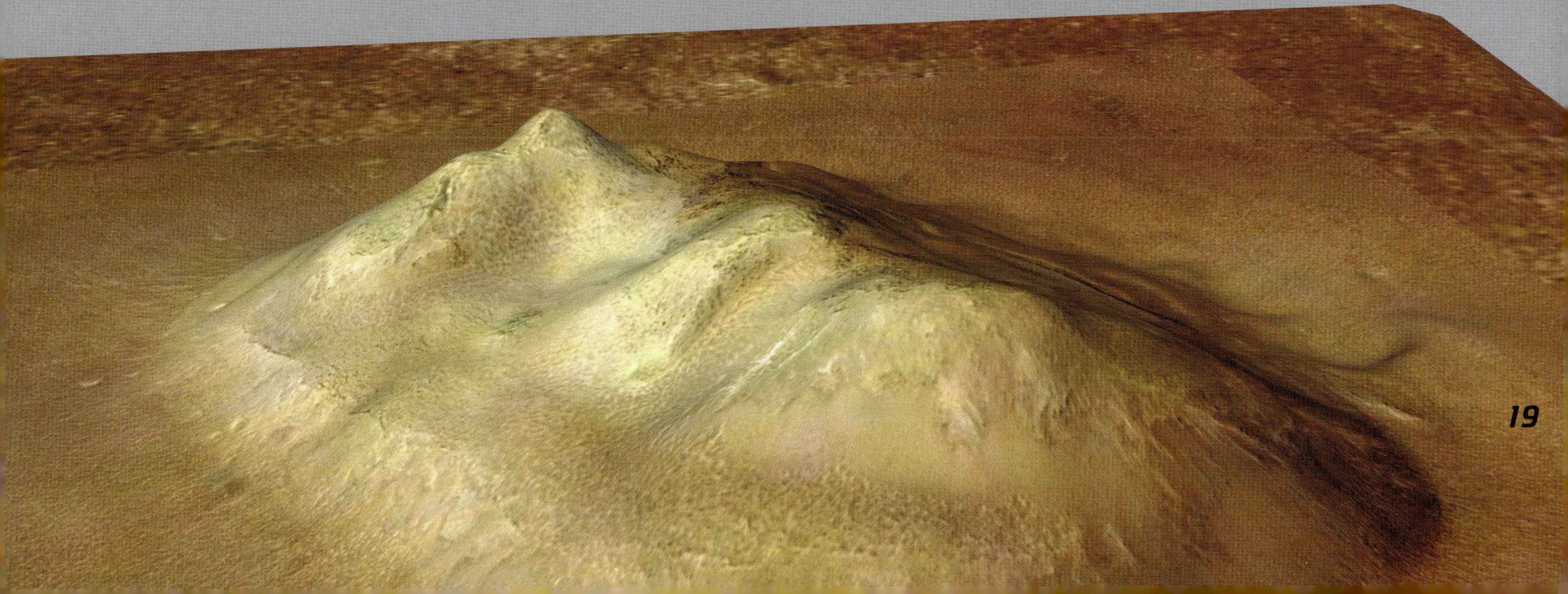

VIKING 2

The second *Viking* probe was also delayed while a suitable landing site was identified. The two orbiters photographed nearly half the planet between them before they found suitable bases. It was all worth it. The pictures and other data that the two landers sent back were remarkably detailed. And they demonstrated, above all, that it was possible to send a spacecraft safely to the surface of Mars.

The two landers observed their surroundings for a Martian year. They recorded the appearance of frost on some rocks. They also sampled the soil. What they found was not encouraging: Mars possesses almost none of the elements, such as carbon and hydrogen, needed to sustain life as we know it.

A CLEAN MACHINE

The *Viking* missions were meant to find out if there was life on Mars. The problem was that the landers could easily carry traces of life with them from Earth. If the scientists weren't careful, they just might mistake a smidgen of Earth bacteria for Mars bacteria. So the landers had to be very, very clean.

Each lander was taken apart after it was built, placed in an oven and baked for forty-eight hours. Then they were reassembled and sealed inside a sterile capsule.

It's almost impossible to make something as complicated as a lander completely sterile. The Martian environment is pretty harsh, however. There was a good chance that any terrestrial microbe that survived the cleaning on Earth would be killed either by radiation or chemicals in the Martian soil when it touched down.

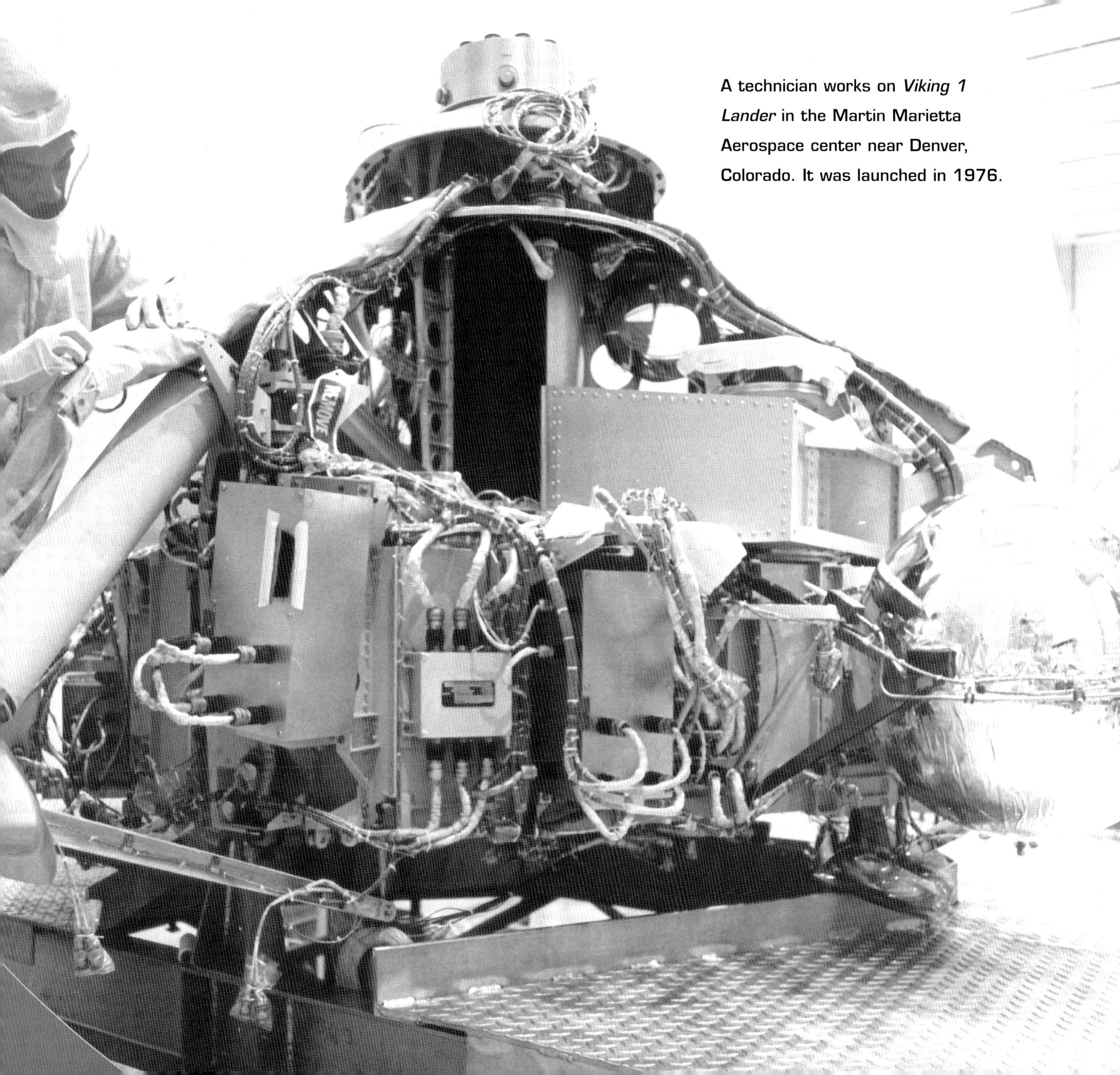

A technician works on *Viking 1 Lander* in the Martin Marietta Aerospace center near Denver, Colorado. It was launched in 1976.

Mars Attacked!

Beginning in the late 1990s, scientists on Earth started launching one mission after another to Mars. Talk about Mars Attacks! If there were critters living on Mars, they might have wondered what hit them . . .

ROBOT SCOUTS

NASA launched *Mars Global Surveyor* in 1996. It was an orbiter, not a lander, but it carried sophisticated cameras and other equipment. It sent back the most detailed pictures of Mars ever seen. It continued to send back pictures and information about Mars for more than ten years. Scientists finally lost contact with the satellite in 2006.

Mars Pathfinder, an orbiter and lander, was launched in the same year as *Global Surveyor*. The rover was named *Sojourner*. It touched down on Mars in July 1997. Unlike the *Viking* landers, which had thrusters to slow them down, the *Pathfinder* lander hit the ground like a beach ball, in a big, gas-filled bag. After impact, the bag was deflated, and *Sojourner* rolled out. The new system worked perfectly the first time.

Mars Odyssey, another orbiter, was launched in 2001. Its main job was to make a careful map of Mars. It also has sent back valuable information about the climate and atmosphere on Mars. It has detected strong evidence that there are large quantities of frozen water mixed into the top layer of Martian soil near its north and south poles. In summer 2009, *Odyssey* still was at work.

ROBOT EXPLORERS

Two more rovers made it to Mars in 2004. *Spirit* and *Opportunity* landed on opposite sides of the planet. They used the same airbag technology that *Pathfinder* used to cushion its arrival, along with a parachute and thrusters.

The landers bounced almost 30 feet (8 m) in the Martian air. Both rovers are much bigger than *Pathfinder*'s was. They each weigh almost 400 lb (180 kg). They're smarter than *Pathfinder* was, too. They carry all kinds of sensitive equipment including microscopes and high-resolution cameras. They have sent back amazing pictures to Earth and more information about the geological history of the planet.

Mars Global Surveyor took this dramatic picture of a canyon wall in Valles Marineris.

A panoramic camera on the big mast of the two rovers, *Spirit* and *Opportunity*, takes wonderful photographs of the Martian landscape.

THE SEARCH GOES ON

Yet another orbiter found its way to Mars in 2006. *Mars Reconnaissance Orbiter* carried even more powerful cameras than the satellites that went before it. It has another function: it will be a communications link between Mars and Earth for future missions. In 2007, several companies and agencies got together to launch the *Phoenix* Mars Mission. *Phoenix* put its rover on the Martian arctic plains to get precise information about the soil there. In the summer of 2009, *Phoenix* was still roaming the Martian pole.

The Martian volcano Olympus Mons is much higher than any mountain on Earth.

The Highs and the Lows

We've scooped, scoped, scraped, scouted, and scrutinized just about every inch of Mars. So what do we know now that we didn't know before? Quite a lot, as it happens.

The northern hemisphere of Mars is different from the southern hemisphere. It's almost as if the halves of two different planets were stuck together.

The southern half is pitted, pockmarked, and cratered. It's much rougher than the northern half, which is generally more sunken and smoother. No one knows for sure why the two halves are so dissimilar. Some scientists think that the northern plains were once the floor of an ancient sea. Others have suggested that they were covered by lava from ancient volcanoes.

POLES APART

Mars has a north and south pole. Both are made of water, ice, and carbon dioxide (kind of like frozen soda water). This icy stuff forms a thin blanket over layers of frozen dust and more ice.

LOOK UP. WAY UP.

There are two enormous plateaus on Mars. Tharsis Plateau is about 4,800 miles long by about 4.2 miles high (8,000 km by 7 km). It is crowned by the greatest volcano in the solar system. Olympus Mons is more than 15 miles (25 km) high—more than three times as high as Earth's Mount Everest. The second great plateau, Elysium, is smaller than Tharsis, but still huge. It's about 1,200 miles long by 3 miles high (2,000 km by 5 km).

LOOK DOWN. WAY DOWN.

The cracks and craters of Mars come in XL and XXL sizes too. Hellas Crater is more than 1,000 miles (1,800 km) in diameter. Argyre is 480 miles (800 km). Most impressive of all is the canyon that follows a path more or less along the Martian equator. Valles Marineris is really a series of connected canyons. Altogether, they stretch about 3,000 miles (5,000 km) and sink to a depth of 4.2 miles (7 km). If Valles Marineris were laid out on North America, it would stretch from New York City to Los Angeles.

Valles Marineris, a system of connected canyons, is far longer and deeper than the Grand Canyon.

THE THREE AGES OF MARS

Noachian Era (1 billion years)

In this era, scientists suspect that Mars was a pretty wild place. Volcanoes spewed out streams of hot dust and molten lava on a planet that was wet, warmer than it is now, and that had a thicker atmosphere.

Hesperian Era (500 million to 1.5 billion years)

Geological activity slowed down while the planet got much colder. Scientists speculate that frozen water trapped underground periodically erupted when the planet was struck by meteorites, causing enormous floods to surge across the surface. Eventually the water migrated beneath the surface to become permafrost.

Amazonian Era (2 to 3 billion years)

Dramatic geological action ended. Mars became dry, as it is now, and its atmosphere much thinner.

Air Today, Gone Tomorrow

When we go to Mars, we either have to take our oxygen with us or manufacture it there. Martian air is not fit to breathe. Martian air is, in fact, about 95 percent carbon dioxide. It also contains traces of nitrogen, oxygen, argon, and water vapor.

Martian air is super thin. The atmospheric pressure on Mars is just about one-hundredth of Earth's. The thin atmosphere means that the boiling point of water, for example, is very low. This is why there is no liquid water on the surface of Mars: any water would instantly turn into vapor.

Mars is about half the size of Earth and it has only about one-tenth of the Earth's mass. Consequently, Martian gravity is comparatively weak.

BOILING MAD

Its atmosphere is so thin that our blood would boil—literally—if we walked on Mars without some kind of protective suit.

RED SKY IN THE MORNING, SAILORS TAKE WARNING

The first *Viking* lander sent back an image of the Martian sky. It was red! Scientists had not expected this. They checked the camera settings to make sure that there was no mistake. They had thought the sky would appear blue, as it does from Earth. After more study, they decided that what they were seeing was fine Martian dust suspended in the atmosphere—not the sky at all.

Martian ice caps may get bigger soon. The average temperature on Mars
has dropped by about two degrees since the mid-1980s. The cause has
nothing to do with greenhouse gases as on Earth. Instead, scientists
speculate that it is related to the gradual darkening of the planet—
meaning that Mars now absorbs more of the sun's heat.

With the approach of autumn in 2008, the
Phoenix Lander photographed blue-gray frost
on Mars' northern plains.

THE FORECAST:
VERY, VERY COLD

Surface temperatures on Mars range from a warm 80°F (27°C)
near the equator in summer to minus 195°F (minus 125°C) at the
poles in winter. The variation is more drastic than we experience
on Earth. Partly this is because there are no oceans on Mars.
Large bodies of water tend to moderate temperature changes.
And partly the variations reflect the long elliptical course of the
planet's orbit, which takes it a long, long way from the Sun.

Hope Springs

What does a Martian look like?

A big blobby critter with eyes like lamps and long, stalky legs?

Or a compact humanoid with green scaly skin and pencil-thin antennae poking from its head?

Probably neither picture is correct.

Chances are, if there's a Martian out there, it looks like nothing we have imagined so far. We know for sure that there is no living creature on Mars that is remotely similar to complex life on Earth.

Life as we know it requires an atmosphere like ours. It requires water in liquid form and organic matter. The atmosphere on Mars is deadly. There is no liquid water on the planet's surface. And, so far at least, there is no conclusive evidence that organic compounds exist there.

So is there life on Mars? There might be, but it isn't a big blobby creature that stares back at us or a lithe alien with its tentacles wrapped around a ray gun.

Soil Sorry

Back in the late 1960s, the two *Viking* landers conducted three identical experiments on a scoopful of Martian soil. In one experiment, nutrients were added to the soil sample. On Earth, the nutrients would have been consumed by microbes. And when they were consumed, they would have released traces of gas. Well, the two landers, each with its separate soil sample, both came up with a positive result when they did this test. Gas—carbon-14—was produced when the nutrients were added to the soil.

The other tests had different results. One was negative, the other was inconclusive. Signs of life? Scientists have argued about this. Most now think that the gas was produced by a chemical rather than a biological reaction. But this conclusion does not end the discussion. There still may be life on Mars.

GOING TO EXTREMES

We know what conditions humans can deal with. Some live in broiling-hot deserts, others in the ice-cold Arctic. We used to think that these environments roughly defined the limits that would support most forms of life.

In recent years scientists have discovered a variety of microbes living in much more extreme habitats than these. Some microbes live in highly acidic hot springs, others at the very bottom of the ocean. Microbes have even been found that can resist radiation. All these microbes are called extremophiles.

Some biologists now believe that the first life on Earth was an extremophile. Perhaps life began in a hot spring, where mineral-rich waters were heated by volcanic activity. Perhaps—some people have dared to speculate—the same process is happening on Mars!

Scientists now speculate that life on Earth began with a tough bacteria—an extremophile such as the sulphur-eating bacteria (bottom) or *sulfolobus archaea* shown here.

Martian Dreams

Wernher von Braun was a German-born rocket enthusiast who worked for the Nazis in World War II. He led the team of scientists that invented the V-2 rocket that was used to bomb Britain and other targets. After the war, he went to work for the American army and later for NASA. He was responsible for developing the *Saturn V* rocket that sent the *Apollo* spaceships to the Moon.

Because of his war work, not everyone admired von Braun. But he was a brilliant scientist and a visionary. Like many other scientists engaged in the space program, he dreamed of sending men to Mars.

Von Braun dreamed big. His plan for Mars, *The Mars Project*, was published in the early 1950s. It called for a massive expenditure of money. Thousands of people would have been involved. It would have taken years and years. But von Braun solved some of the basic problems of interplanetary travel.

The toughest problem is that humans have to take everything with them to Mars. Everything includes food, water, fuel, and even air to breathe. Water and fuel, especially, are bulky and heavy. Everything also includes the astronauts' home. The first travelers to Mars will be away from Earth for at least a couple of years. They need to take their living and working quarters with them. There is no way to pack all these components onto the top of a single rocket. There is no rocket big enough to throw such a load of stuff out of Earth's gravitational field and into space.

Von Braun's solution? Build a space station in low Earth orbit and then launch the mission to Mars from there.

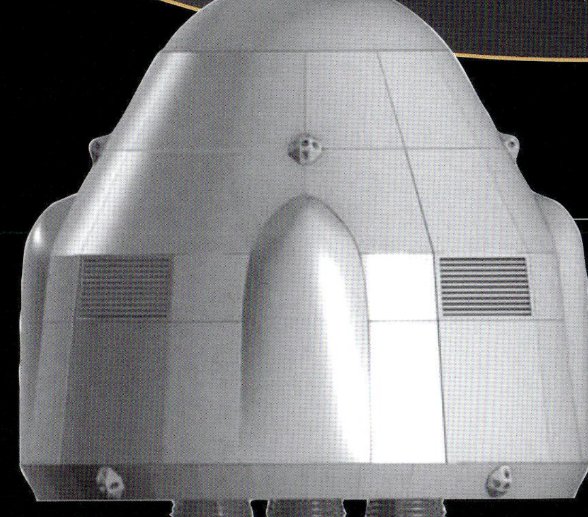

HAPPY LANDINGS

● ● ● ● ● ● ● ● ● ● ● ● ● ● ● ●

Von Braun proposed sending ten 3,630 metric-ton (4,000-ton) spaceships and seventy crewmembers to Mars. Seven of these ships would be stationed in Mars orbit. Three of them would be designed to land on Mars. The first of the three would land on one of the polar ice caps. (Von Braun believed that only the polar ice would be smooth enough to land on.) The crew would then make the incredibly difficult 4,000-mile (6,400-km) trek to a site near the equator where they would build a landing strip for the two other landers. Only after the crews of all three spaceships met up would the work of exploration and investigation begin.

Artist's conception of the Earth Return Vehicle (ERV) that will bring the first Mars explorers back to Earth.

Wernher von Braun

31

The Junkyard *Special*

Robert Zubrin is a modern visionary in the Wernher von Braun tradition. He is determined to make the Mars dream a reality. He believes that von Braun's plan is too big and expensive. Zubrin's plan is meant to be cheaper and more practical.

Zubrin argues that we don't need to invent a new rocket. We can use the technology we already possess to put together a suitable vehicle. His Mars rocket, called *Ares*, will be based on the big *Saturn V* but uses engines and boosters that were developed for the space shuttle. Because it uses bits and pieces that already exist, he jokes that *Ares* will be a "junkyard special."

LIVING OFF THE LAND

Zubrin believes that the astronauts who go to Mars can learn a thing or two from the explorers who first traveled to the North Pole on Earth. The first explorers took everything they thought they would need with them. They packed woolen underwear to keep them warm and food preserved in tins to sustain them. But wool wasn't warm enough and the tins, which were sealed with lead, may have poisoned them. Later expeditions learned from the Inuit how to live off the land. They wore caribou skin instead of woolies and ate seal meat instead of tinned beef.

Zubrin wants the first explorers on Mars to be like the later explorers in the Arctic. He wants them to use the resources they find there.

Following his plan, an unmanned spaceship called an Earth Return Vehicle (ERV), carrying a nuclear reactor and a chemical plant, is sent to Mars. The nuclear reactor provides power for the chemical plant. When it lands on Mars, the chemical plant goes to work. It sucks in Martian air, which is mostly carbon dioxide, adds hydrogen to start a chemical reaction, and produces methane. The methane is stored away to be used as rocket fuel for the journey home. By the end of a few months, by Zubrin's calculations, there will be more than enough fuel for the ERV and its passengers' ride home.

Zubrin also believes that the astronauts on Mars are likely to find water beneath the planet's surface. When they do, they can use the water both for drinking and to irrigate farms enclosed in greenhouses. In addition, the hydrogen in water (water, or H_2O, is partly hydrogen) can be used in the chemical plant to make more rocket fuel.

By using Martian resources, the Martian explorers not only reduce the cost and complexity of their voyage, but they also increase their chances of surviving.

A Moving Target

Robert Zubrin's plan comes as close to saying, *Hey, let's just go!* as it's possible to imagine. It works like this:

First, an unmanned base is established on Mars, complete with an ERV fueled and ready for take-off. Next, two *Ares* rockets are launched from Earth. One carries a crew in a habitation module or *hab*. The other is an unmanned ERV that will be landed and fueled on Mars, just as the first one was. It awaits the arrival of a second human crew.

The first crew lands and spends sixteen months on Mars. (Zubrin feels that, having come so far, the Mars pioneers will want to stay for a while.) They will be busy with many scientific tasks. Drilling for water will be a priority. The second crew will not leave Earth until the first crew has shown that the base is habitable and functioning. Gradually, however, more crews will arrive as the first ones leave and the settlement on Mars will be expanded.

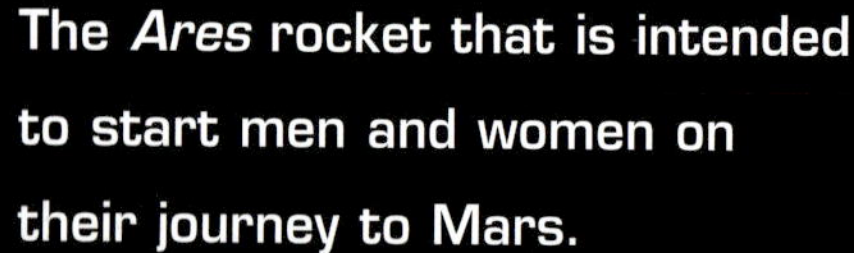

The *Ares* rocket that is intended to start men and women on their journey to Mars.

VENUS FLY TRIP

Suppose your home and school both won't stay still. They are constantly on the move. And then suppose that they are not just a few miles, but *millions* of miles apart. Now you have to find the quickest route between them. It's a good thing getting to school isn't really this tough!

Working out how to get to Mars poses a similar challenge. Both Earth and Mars are in different orbits circling the Sun. The distance between them varies from about 60 million miles (100 million km) to about 235 million miles (380 million km). Surprisingly, the shortest route between them may not be the best one.

A number of scientists have proposed sending a manned Mars mission on a path that takes it close to Venus. As the spacecraft approaches the planet, it is pulled into Venus' gravitational field. In effect, the spacecraft gets a free ride—*a gravity boost*—for part of its journey. This means that it needs less fuel than it would need if it flew entirely under its own power—a good thing. Anything that makes the spacecraft's cargo lighter is an advantage.

Do You Have the Right Stuff?

1. Are you good at science or math?

Almost all astronauts are engineers or scientists.

2. Do you enjoy doing school work?

You had better enjoy it! Most astronauts have more than one college degree.

3. Would you rather spend time with friends than be alone?

You are going to spend a lot of time with other people on the voyage to Mars. You have to be able to get along with them.

4. Do you play a team sport?

The ability to work with others to achieve a common goal is essential.

5. Do you enjoy physical activity?

This is not a job for a couch potato! You have to be physically fit.

6. Is it important to you to win at games?

The space program needs people who have both a healthy competitive instinct and a generous nature. You have to want to be the best you can be.

7. Are you a member of a club?

It helps if you have outside interests. You need to know a lot of different things.

8. Are you the president of a club or captain of a sports team?

The space program is looking for leaders.

9. Are you or have you been a Scout?

A high percentage of astronauts have participated in the Scout movement.

10. Would you describe yourself as well-organized and decisive?

The space program needs people who get things done.

Neil Armstrong walked on the Moon for two-and-a-half hours in 1969. Without question, he had the "right stuff."

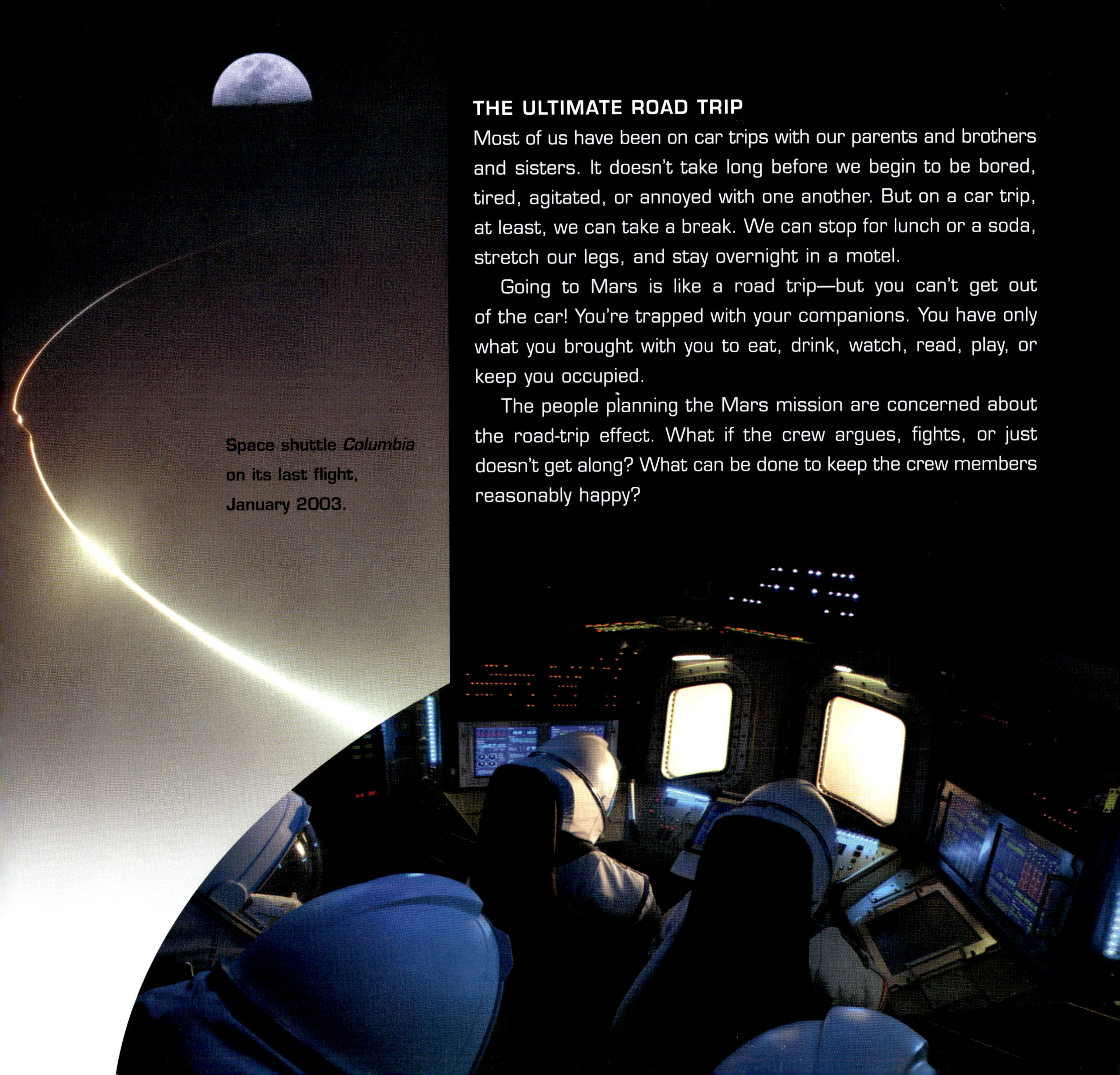

THE ULTIMATE ROAD TRIP

Most of us have been on car trips with our parents and brothers and sisters. It doesn't take long before we begin to be bored, tired, agitated, or annoyed with one another. But on a car trip, at least, we can take a break. We can stop for lunch or a soda, stretch our legs, and stay overnight in a motel.

Going to Mars is like a road trip—but you can't get out of the car! You're trapped with your companions. You have only what you brought with you to eat, drink, watch, read, play, or keep you occupied.

The people planning the Mars mission are concerned about the road-trip effect. What if the crew argues, fights, or just doesn't get along? What can be done to keep the crew members reasonably happy?

Away from Home

The Mars Habitation Module, or *hab*, will be both home and workplace for the first travelers to Mars. It will be as comfortable as planners can make it. Just how comfortable depends, in part, on how grand the expedition is. Will it be as Wernher von Braun envisioned it: a great big station built in space? Or will it be the stripped-down module that Robert Zubrin described?

WEIGHTY MATTERS

Some planners want to create artificial gravity in the spaceship on its way to Mars. They believe that they can do this by designing the hab so that it spins around a central axis. Think of the feeling you get when you ride on a carousel: if you don't hang on, the *centripetal force* generated by the rapid circular motion tends to pull you towards the edge. This same force could be used to simulate gravity in the hab.

Artificial gravity would make the hab more like Earth. It might also be healthier. The men and women on board the *International Space Station* have lived without gravity for months at a time. Thanks to them, we now know what weightlessness can do to the human body.

When we are on Earth, our body fluids tend to flow towards our feet. Our muscles push the fluids back up towards the heart. Without gravity, however, these fluids tend to rise. Muscles, especially in the legs, don't have to work as hard when

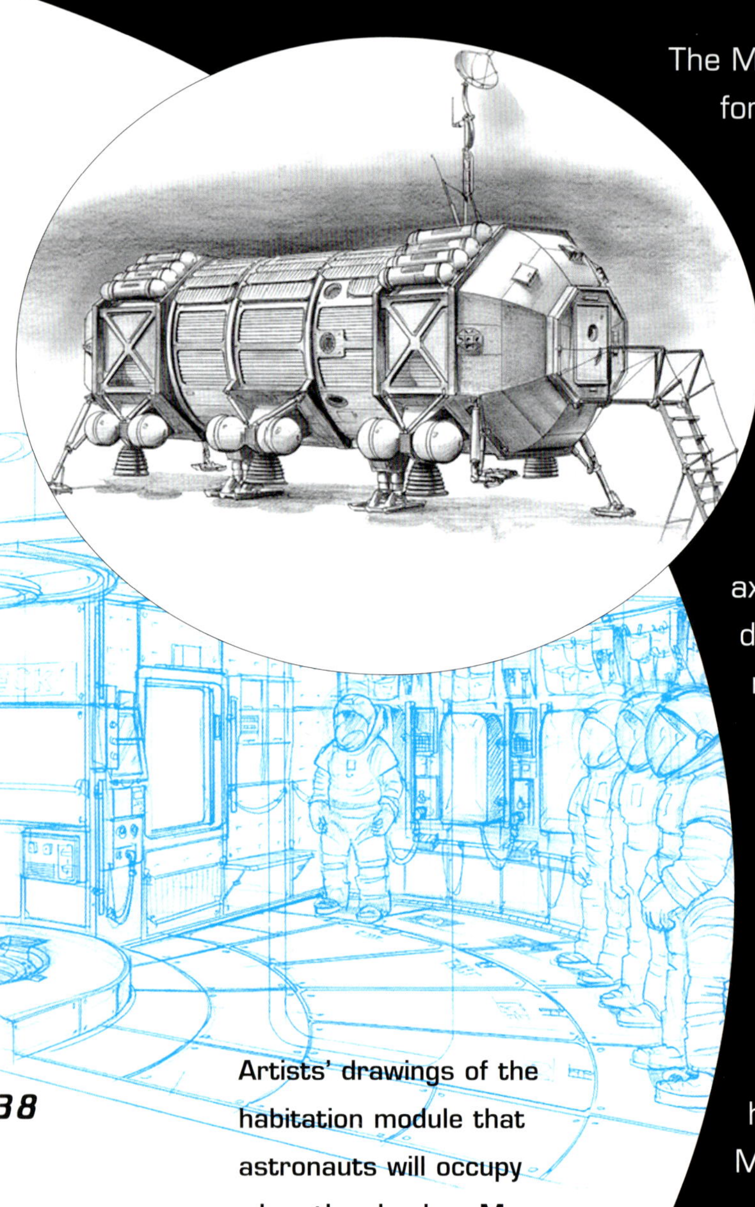

Artists' drawings of the habitation module that astronauts will occupy when they land on Mars.

they don't have to fight against gravity. Consequently, they tend to weaken and atrophy. Everyone on the *International Space Station* follows a rigorous exercise program to keep fit.

For some reason, not yet fully understood, bones tend to lose their mineral content when the body is weightless for a long time. Crew members may be more likely to suffer a bone fracture after a long trip.

The case for building artificial gravity into plans for the spacecraft is a strong one. There are arguments against it, however. For one thing, the ship itself will be more complicated and expensive to build with artificial gravity. For another, it will require more fuel to keep it spinning. And finally, it may make communications and navigation more difficult.

SPACE-CRAFTY FELLOWS

● ● ● ● ● ● ● ● ● ● ● ●

The American astronaut Michael Collins noticed that his colleagues on *Apollo 11* looked different: "Because of the lack of gravity . . . wrinkles disappeared and their eyes looked squinty and crafty. The three of us also grew by an inch or two with no gravity to compress the space between our vertebrae."

How Many Pairs of Socks Will I Need?

WHAT'S FOR DINNER?

The average man consumes 3 lb (1.36 kg) of food a day, the average woman somewhat less. Over the course of a twenty-two month journey to and from Mars (the time proposed by one writer, astronaut Michael Collins) the total supply of food for each crewmember would be 1,980 lb (900 kg). That's a lot of food.

The space shuttle provided hot and cold food, a pantry, an oven, and cleaning facilities. Seventy different food items and twenty different beverages were available.

Much of the food supply must be partly or completely dehydrated. That is to say, its water content is reduced, as it is in powdered soup mix or fruit leather. Some foods, like crackers or nuts, can be carried on board unchanged. Others may have to be preserved by freeze-drying or irradiation. Crumbly foods, including bread, are ruled out: the mess they make can cause real problems on a spacecraft. Space shuttle crews ate tortillas instead of bread.

The Mars-bound astronauts will savor an occasional morsel of fresh fruit or vegetables. They'll have to grow it themselves. NASA planners believe peanuts, potatoes, tomatoes, and soy may be among the crops grown on the way to Mars.

Thanks to the *Progress* supply vehicle, astronauts on the *International Space Station* (*ISS*) never wash their laundry.

The crew on board the *ISS* demonstrate their food storage system.

THIS WATER LOOKS FAMILIAR

All waste water—including dish and shower water, water vapor, and even urine—will be purified and recycled to be used again for everything from drinking to doing the laundry.

SUDS FOR DUDS

Astronauts on the space shuttle wear a different outfit every day. They bring their laundry back to Earth. Crewmembers on the *International Space Station* change their underwear and socks every other day. The rest of their outfits are made to last a little longer. Nothing is laundered in space. Dirty clothes are packed into bags and loaded onto the re-supply vehicle, which is then pointed towards Earth. It and the laundry burn up on entering the Earth's atmosphere.

The journey to Mars is a long one. Crewmembers will have to wash what they wear. How many pairs of socks will they need? About as many as they would have on Earth.

41

Risky *Business*

Scientists joke about the Great Galactic Ghoul but the danger that inspired the joke is real. A lot of missions bound for Mars have failed. A lot of things can go wrong.

DON'T FORGET YOUR UMBRELLA!

Not for the rain—but for the *radiation*.

Our atmosphere protects life on Earth by screening out most cosmic and solar radiation. Travelers in space are much more vulnerable. Solar flares—periodic eruptions on the surface of the Sun that release bursts of radiation—could be fatal to the astronauts on their way to Mars. The *Skylab*, which orbited Earth in the 1970s, carried a thin shield that opened up like a space umbrella to help protect it from radiation. We know now that polyethylene—the same plastic found in soda bottles—can be used in construction of the spacecraft to make an effective shield.

BLAST OFF

Basically, a rocket launch is a huge, controlled explosion. *Saturn* set off an explosion that was incredibly powerful. The *Ares* rockets being developed for the Mars mission are just as powerful. And a brave crew of astronauts will be sitting on top of that roaring conflagration at take-off.

Yes, it is dangerous.

DO YOU SMELL SOMETHING?

Various kinds of mechanical breakdown, electronic failure, and fire are all possible hazards of space travel. The Mars crew is on its own—they can't call the fire department in an emergency!

In 2004, astronauts on board the *International Space Station* discovered that they had a toxic leak in the ventilation system. The Elektron unit, which makes breathable oxygen by separating hydrogen from oxygen atoms in water, had broken down. The crew was able to track down the leak and fix it before anyone was harmed. The incident showed again how every single piece of the spacecraft has to function efficiently.

In spite of all the dangers, the astronauts on their way to Mars will wake up every day knowing that they are true pioneers. They can look out the portholes and see the universe as no one has ever seen it. And they can look forward to an incredible adventure as the first humans to set up camp on another planet!

KA-BOOM!

● ● ● ● ● ● ● ● ●

The first stage of the *Saturn V* rocket, which propelled the *Apollo* space capsules into orbit, carried more than 2,200 tons (2,000 metric tons) of kerosene fuel and liquid oxygen oxidizer for its five main rocket engines. Once they were ignited, each of those engines consumed 15 tons (13.6 metric tons) of propellant per second to develop a total of 3,700 tons (3,400 metric tons) of thrust at lift-off.

The First
Martians

Nothing could have prepared you for this!

With other astronauts, you spent time in Earth's Arctic. The experience was meant to give you a glimpse of the wide-open, rugged terrain you would find on Mars. It didn't come close! The landscape on Mars is harsher, harder, higher, deeper, and more barren than anything on Earth.

You have ridden the Mars buggy—the electric jeep officially known as the *surface exploration vehicle* (SEV)—and glimpsed the almost unimaginable immensity of a Martian mountain. You have peered down a chasm so sharp and deep that you wondered if it had opened up, in some great tectonic shakeup, just hours before your arrival. (But you know it didn't. It's billions of years old.) You have taken shelter from a Martian dust storm. This, at least, you had expected. And yet, you could not have anticipated its all-encompassing ferocity. No storm on Earth covers the Earth entirely. This thunderous, swirling, blinding mass of red grit blanketed the hab for three Martian weeks. For all that time, the sound of the windblown particles lashing the outer walls in waves of rising and falling intensity never stopped for a second. And yet you know, by Martian standards, the storm was not unusual.

You are almost accustomed to the color.

For the first few days on Mars, you kept wiping the back of your gloved hand across your visor. You were trying instinctively to wipe away the dust that made everything look red. But no amount of wiping erased it. You still find this unsettling. Red is not a restful color.

Since the storm ended, however, you have been either working or sleeping. You have had very little time—and no energy—to think about what you are going through. There will be more than enough time (months!) to think on the journey home. For now, you have umpteen experiments to monitor. And you have to take your turn working on the drill rig, probing the Martian crust for water. Yes, there's more than enough to do, and so little time.

You are a Martian now. It is your planet. You claimed it years ago when you first found it flickering ominously in Earth's night sky. Now Mars has claimed you.

The crew strikes water! But because the air pressure is low and the temperature so cold, the water turns instantly into snow and ice—a bracing shower for the men and women operating the drill.

Further Reading and Other Resources

Dana Berry. *Race to Mars*. Barron's / Madison Press, 2007.

Joseph M. Boyce. *The Smithsonian Book of Mars*. Washington: Smithsonian Institution Press, 2002.

Andrew Chaikin. *A Passion for Mars: Intrepid Explorers of the Red Planet*. New York: Abrams, 2008.

Michael Collins. *Mission to Mars*. New York: Grove Weidenfeld, 1990.

Robert Zubrin. *The Case for Mars: The Plan to Settle the Red Planet and Why We Must*. New York: The Free Press, 1996.

The internet has superb resources for further exploration. The following organizations represent some useful entry points. Enter the suggested key words in your internet search engine.

The National Aeronautical and Space Agency (NASA) is the primary resource for all things space-related. Enter the key word "NASA."

The Euopean Space Agency (ESA) conducts research and has launched a variety of space missions. Enter the key words "European Space Agency."

The recently repaired Hubble Telescope has sent amazing photographs to Earth from space. Enter the key words "Hubble Telescope."

The Lowell Observatory web site offers insight into the past and present state of astronomy. Enter the key words "Lowell Observatory."

The Smithsonian Institution maintains the National Aeronautical and Space Museum. Enter the key words "National Aeronautical Space Museum."

A thought-provoking exhibition called "Facing Mars" was developed by the Ontario Science Centre in consultation with leading experts in the field of space exploration, to both educate and entertain audiences. Enter the key words "Facing Mars."

Acknowledgements

This book would scarcely be possible were it not for the public relations division of NASA and its superb internet resources. The European Space Agency was similarly helpful. We are also grateful to Quickplay Media Inc., Discovery Channel Canada and Galafilm. "Race To Mars" and the documentary "Mars Rising" were both produced by Galafilm with the financial participation of Discovery Channel Canada, the Canadian Television Fund, the Quebec Film and Television Tax Credit, the Canada Film or Video Production Tax Credit, 13 Production, ARTE and Discovery Networks International. The interactive "Race to Mars" project was produced by Quickplay Media Inc. with the financial participation of the Quebecor Fund, the Bell Broadcast and New Media Fund, Telefilm Canada Administrator of the Canada New Media Fund funded by the department of Canadian Heritage and the Ontario Media Development Corporation's Interactive Digital Media Fund. Author, artist, and Mars enthusiast, Dana Berry, kindly supplied images and sage advice.

ESA	European Space Agency
Galafilm	Galafilm Productions (XII) Inc.
NASA	National Aeronautics and Space Administration
NASA/JPL/Caltech	National Aeronautics and Space Administration/ Jet Propulsion Laboratory/California Institute of Technology
NASA/JPL/GSFC	National Aeronautics and Space Administration/ Jet Propulsion Laboratory/Goddard Space Flight Center
NASA/JPL/Malin	National Aeronautics and Space Administration/ Jet Propulsion Laboratory/Malin Science Systems
NASA/JPL/USGS	National Aeronautics and Space Administration/ Jet Propulsion Laboratory/United States Geological Survey
NASA/NSSDC	National Aeronautics and Space Administration/ National Space Science Data Center

Endpapers: Detlev Van Ravenswaay/
   Science Photo Library
1: Galafilm
2: (Left) NASA; (Background) NASA/JPL/Malin
3: (Top) Galafilm; (Right) NASA
4: iStock International Inc.
5: Corbis
6: Artwork by Dana Berry
7: Getty; (Background) NASA/JPL/GSFC
8: Corbis
8-9: Collection of J. Webb
9: (Right) Corbis
10: Corbis; (Background) NASA/JPL/Caltech
11: NASA/NSSDC; (3 insets) Collection of
   J. Webb; (Background) NASA/JPL/Caltech
12: Corbis
12–13: (Background) Artwork by Dana Berry
   (background sky from 1000skies.com
   via Sky Works Digital)
14: NASA
15: NASA
16: (Left) Getty; (Right) NASA
17–19: NASA
20: (Top) Getty; (Bottom) iStock International Inc.

21–23: NASA
24: (Left) NASA/Science Photo Library;
   (Right) Getty
25: NASA/JPL/USGS
26: ESA/DLR/FU Berlin
27: ASA/JPL/Caltech/University of Arizona/
   Texas A&M University
28: (Top) NASA; (Bottom) iStock International Inc.
29: (Top) Eye of Science/Science Photo Library;
   (Bottom) Pasieka/Science Photo Library
30: Getty
31: Galafilm
32: NASA/JPL/Caltech
33: Galafilm
34: NASA
35: Galafilm
36: NASA
37–38: Galafilm
39: (Inset) NASA; (Bottom) Galafilm
40: NASA
41: (Top) NASA; (Bottom) iStock International Inc.
42: NASA
43–46: Galafilm
47: NASA/JPL/Malin

Text, Design and Compilation © 2009 Madison Press Books

First published in Canada in 2009 by
Scholastic Canada Ltd.
604 King Street West
Toronto, Ontaro
M5V 1E1

10 9 8 7 6 5 4 3 2 1

Library and Archives Canada Cataloguing in Publication

Webb, Jonathan, 1950–
Journey to Mars: quest for the red planet / Jonathan Webb

ISBN 978-1-4431-0009-0

1. Mars (Planet)–Juvenile literature. 2. Mars (Planet)–
Geology–Juvenile literature. I. Title.

QB641.W43 2009   j523.43   C2009-904374-2

Printed in China

Produced by
Madison Press Books
1000 Yonge Street, Suite 303
Toronto, Ontario M4W 2K2

Editor
Jonathan Webb

Editorial Assistant
Hannah Draper

Production Editor
Mollie Wilkins

Art Director
Diana Sullada

Printed by
Oceanic Graphic Printers